TOP SECRET

**SECRET INTELLIGENCE
FIELD MANUAL –

STRATEGIC SERVICES**
(Provisional)

Prepared under direction of
The Director of Strategic Services

TOP SECRET

OSS REPRODUCTION BRANCH

TOP SECRET

SECRET INTELLIGENCE FIELD MANUAL

– STRATEGIC SERVICES
(Provisional)

Strategic Services Field Manual No. 5

TOP SECRET

TOP SECRET

Office of Strategic Services
Washington, D. C.
22 March 1944

This Secret Intelligence Field Manual — Strategic Services (Provisional) is made available for the information and guidance of selected personnel and will be used as the basic doctrine for Strategic Services training for this subject.

The contents of this Manual should be carefully controlled and should not be allowed to come into unauthorized hands. The Manual should not be taken to advance bases.

AR 380-5, 15 March 1944, pertaining to the handling of secret documents, will be complied with in the handling of this Manual.

William J Donovan

William J. Donovan
Director

TABLE OF CONTENTS

SECTION I — INTRODUCTION

1. SCOPE AND PURPOSE OF MANUAL . . 1
2. DEFINITIONS 1
3. FUNCTIONS OF THE SECRET INTELLIGENCE BRANCH 3

SECTION II — ORGANIZATION

4. ORGANIZATION IN WASHINGTON . . . 5
5. ORGANIZATION AT OSS FIELD BASES . . 6
6. ORGANIZATION WITHIN NEUTRAL COUNTRIES 6
7. ORGANIZATION OF OPERATIVES AND AGENTS IN THE FIELD 7

SECTION III — PERSONNEL

8. ORGANIZATION FOR RECRUITMENT . . 7
9. SOURCES FOR THE RECRUITMENT OF PERSONNEL 7
10. TYPES OF PERSONNEL REQUIRED . . . 8

SECTION IV — TRAINING

11. ORGANIZATION FOR TRAINING . . . 8
12. SCOPE OF TRAINING 8
13. TRAINING OBJECTIVES 9

SECTION V — METHODS OF OPERATIONS

14. GENERAL 10
15. SECURITY 10
16. COVER 11
17. COMMUNICATIONS 14
18. ARRIVAL AND DEPARTURE 15
19. ESTABLISHING SOURCES OF INFORMATION 16
20. REMUNERATION OF AGENTS 19
21. SPECIALIZATION OF AGENTS 20
22. DOUBLE AGENTS 20

23. SEVERING CONNECTIONS WITH AGENTS 21
24. RELATIONS WITH UNDERGROUND
     GROUPS 21
25. ASCERTAINING PUBLIC OPINION . . . 21
26. PENETRATING AN ENEMY OR ENEMY-OC-
     CUPIED COUNTRY FROM NEUTRAL
     TERRITORY 22
27. REPORTS 23
28. RECORDS AND DOCUMENTS 24
29. RELATION BETWEEN SI DESKS, WASHING-
     TON, AND SI STAFF IN THE FIELD . 26

SECTION VI — TYPES OF INFORMATION REQUIRED

31. GENERAL 27
32. MILITARY INFORMATION 27
33. NAVAL INFORMATION 28
34. ECONOMIC INFORMATION 29
35. POLITICAL INFORMATION 30
36. PSYCHOLOGICAL INFORMATION . . . 31

SECTION VII — COOPERATION OF SI WITHIN OSS
AND WITH OTHER ORGANIZATIONS

37. GENERAL 32
38. COOPERATION WITH OTHER BRANCHES
     OF OSS 32
39. COOPERATION WITH THE ARMED FORCES 33
40. COOPERATION WITH THE DEPARTMENT
     OF STATE 34
41. COOPERATION WITH SIMILAR AGENCIES
     OF ALLIED NATIONS 34

SECTION VIII — PLANNING

42. GENERAL 35
43. PROGRAMS 35
44. IMPLEMENTATION STUDIES 36
45. CHECK LIST 36

TOP SECRET

SECRET INTELLIGENCE FIELD MANUAL
STRATEGIC SERVICES

(Provisional)

SECTION I — INTRODUCTION

1. *SCOPE AND PURPOSE OF MANUAL*

This manual sets forth the operational principles, methods, and organization of Secret Intelligence as a part of Strategic Services activities, exclusive of that obtained by counter-espionage methods, which is covered by a separate manual. Its purpose is to provide guidance to authorized SS personnel engaged in operational planning and training in Washington and at field bases. In view of its highly secret nature, this manual will be given a very limited distribution.

2. *DEFINITIONS*

a. OVER-ALL PROGRAM FOR STRATEGIC SERVICES ACTIVITIES—a collection of objectives, in order of priority (importance) within a theater or area.

b. OBJECTIVE—a main or controlling goal for accomplishment within a theater or area by Strategic Services as set forth in an Over-All Program.

c. SPECIAL PROGRAM FOR STRATEGIC SERVICES ACTIVITIES—a statement setting forth the detailed missions assigned to one or more Strategic Services branches, designed to accomplish a given objective, together with a summary of the situation and the general methods of accomplishment of the assigned missions.

d. MISSION—a statement of purpose set forth in a special program for the accomplishment of a given objective.

e. OPERATIONAL PLAN—an amplification or elaboration of a special program, containing the details and means of carrying out the specified activities.

f. TASK—a detailed operation, usually planned in the field, which contributes toward the accomplishment of a mission.

g. TARGET—a place, establishment, group, or individual toward which activities or operations are directed.

h. THE FIELD—all areas outside of the United States in which Strategic Services activities take place.

i. FIELD BASE—an OSS headquarters in the field, designated by the name of the city in which it is established, e.g., OSS Field Base, Cairo.

j. ADVANCED OR SUB-BASE—an additional base established by and responsible to an OSS field base.

k. OPERATIVE—an individual employed by and responsible to the OSS and assigned under special programs to field activity.

l. AGENT—an individual recruited in the field who is employed and directed by an OSS operative or by a field or sub-base.

m. SUB-AGENT—an individual not a regular member of OSS who is employed and directed by an agent in the field. Sub-agents may be paid or they may be volunteers.

n. INFORMANT—an individual who, knowingly or unknowingly, gives information to an OSS operative, agent, or sub-agent.

o. COVER—an open status, assumed or bona fide, which serves to conceal the secret activities of an operative or agent.

p. CUTOUT—a person who forms a communicating link between two individuals, for security purposes.

q. RESISTANCE GROUPS—individuals associated together in enemy-held territory to injure the enemy by any or all means short of military operations, e.g., by sabotage, espionage, non-cooperation.

r. GUERRILLAS—an organized band of individuals in enemy-held territory, indefinite as to number, which conducts against the enemy irregular operations including those of a military or quasi-military nature.

3. FUNCTIONS OF THE SECRET INTELLIGENCE BRANCH

<u>a</u>. The principal function of the Secret Intelligence Branch is to collect and evaluate secret intelligence and to disseminate such intelligence to appropriate branches of OSS and to military and other authorized agencies. Supplementary functions are: to establish and maintain direct liaison with Allied secret intelligence agencies; and to obtain information from underground groups by direct contact or other means.

(1) *Collection of information*

Information is collected in neutral, enemy, and enemy-occupied countries, outside of the Western Hemisphere, by secret intelligence operatives and agents working under cover. This information is obtained by personal observation, through strategically placed informants, or by other means available. Information is also collected in Allied countries through contact with Allied secret intelligence agencies and representatives of underground or other groups and from individuals who have special knowledge.

(2) *Evaluation of information*

(a) Information is evaluated both as to the reliability of the source and as to the truth, credibility, or probability of the information itself. The following rating scale is used in evaluating the source:

- A — Completely reliable
- B — Usually reliable
- C — Fairly reliable
- D — Not usually reliable
- E — Unreliable
- F — Untried

TOP SECRET

(b) The following rating scale is used in evaluating the truth, credibility, or probability of the information:

1 — Report confirmed by other sources
2 — Probably true
3 — Possibly true
4 — Doubtful
5 — Improbable
0 — Truth cannot be judged

Thus a report rated A-2 would be a probably true report coming from a completely reliable source.

(c) In Washington, the responsibility for the evaluation of information is lodged in the SI Reporting Board. In the field, Reports Officers perform this function. So far as the evaluation of the *source* of material is concerned, the field offices and the desk heads, through the maintenance of records on operatives and agents, are able to furnish the reporting officials with information from which reasonable conclusions may be drawn. As to the presumptive reliability of the *content* of reports, the operatives and the field offices contribute their opinion, the geographic desk heads add whatever comment they may be in a position to make, and the reporting officials check the information against their own records and knowledge and against information available in other branches of OSS, particularly R&A, or in other government agencies.

(3) *Dissemination of intelligence*

(a) SI disseminates intelligence to the other branches of OSS, and selected intelligence to the Military Intelligence Division (MID), the Office of Naval Intelligence (ONI), Air Intelligence (A-2), Joint Intelligence Committee, the State Department, other authorized U. S. government agencies, and to the designated authorities of Allied governments.

(b) Secret intelligence is also disseminated from field bases either directly by the Reports Officer or through the Joint Intelligence Collection Agencies (JICA)* in the Theaters of Operations where such agencies have been established.

(c) Dissemination of secret intelligence is the function of the Reports Officer at an OSS field base and of the SI Reporting Board at Washington. The SI desk heads may suggest the dissemination to be given a report.

(d) In general, operatives should not attempt to disseminate intelligence within the actual area of operations, both for reasons of security and for lack of ability properly to evaluate.

SECTION II — ORGANIZATION

4. *ORGANIZATION IN WASHINGTON*

a. The SI Branch is one of the intelligence branches, under the general supervision of the Deputy Director, Intelligence Services, OSS.

b. The Chief, SI Branch, is charged with the responsibility of carrying out the functions of the Secret Intelligence Branch. Deputy chiefs are responsible to the Chief, SI Branch, for the supervision of activities within certain broad geographic areas. Under the deputy chiefs, there are section chiefs responsible for all SI activities within smaller geographic areas or theaters or neutral countries. Desk heads are responsible to section chiefs for SI activities in smaller areas which are subdivisions of the section areas.

c. In addition to the organization along geographic lines for SI activities, there are functional sections covering special activities which cut across geographic lines. The section chiefs for these functional sections,

---

* The functions of the Joint Intelligence Collection Agencies are to: (1) obtain all information within the Theater which is desired by the War and Navy Departments; (2) coordinate, consolidate, and evaluate such information and forward it to Washington by the most expeditious means for dissemination among the interested agencies; and .') furnish the Theater Commander with such of this information as he desires.

TOP SECRET

while directly responsible to the Chief, SI, for their specialized activities, work in close cooperation with the geographical section chiefs.

<u>d</u>. The Executive Officer, SI Branch, is a general executive assistant to aid the chief in the performance of all duties assigned to him.

<u>e</u>. Special staff officers are responsible to the chief for planning, personnel, and administration, as follows:

(1) The Planning Officer has the duty of coordinating the plans of the various sections and desks within SI, with the over-all and special plans for Strategic Services activities.

(2) The Personnel Officer is responsible for the recruiting and training of all SI personnel (except clerical).

(3) The Administrative Officer coordinates all administrative activities within SI having to do with procurement of supplies, budget and finance, special funds, administrative services, civilian clerical personnel, communications and medical services.

<u>f</u>. The Reporting Board is responsible for the evaluation and dissemination of intelligence.

5. *ORGANIZATION AT OSS FIELD BASES*

Organization at OSS field bases varies according to the function of the base and the number of personnel on the staff. In general it follows the pattern of the organization at Washington, with this exception: All the activities of a field base within a theater of operations are under the control and direction of the theater commander concerned who exercises this direction and control through the Strategic Services Officer.

6. *ORGANIZATION WITHIN NEUTRAL COUNTRIES*

SI operates as a section of the OSS mission under the direct control and supervision of the Chief of OSS Mission.

TOP SECRET

7. ORGANIZATION OF OPERATIVES AND AGENTS IN THE FIELD

Local conditions, the mission to be performed, and the availability of personnel will determine the number and development of operatives and agents in the field. It may vary from one SI operative or agent to an organization consisting of numerous personnel comprising one or more networks.

SECTION III — PERSONNEL

8. ORGANIZATION FOR RECRUITMENT

a. Section chiefs and desk heads are responsible for initiating and following up the recruiting or transfer of civilian personnel (except clerical), and for initiating the transfer of military personnel to OSS. They are also responsible for the final selection.

b. The Personnel Officer, SI, assists the section chiefs and the desk heads in filling their personnel requirements, and channels all requests for recruitment of military, naval, and civilian personnel through the Personnel Procurement Branch, OSS. In addition, he assists in coordinating within the SI Branch all matters relating to the procurement of personnel other than clerical or stenographic.

c. All personnel for SI activities must be approved by the Security Officer, OSS.

d. Personnel for field bases are generally recruited in Washington. Where recruiting is done locally, it is the responsibility of the SI section head of the OSS field base, under the control and direction of the Strategic Services Officer.

e. The recruiting of agents, sub-agents, and informants in the field may be done by an operative or by a desk head at a field base. (See Section V, paragraph 18, for a discussion of this matter.)

9. SOURCES FOR THE RECRUITMENT OF PERSONNEL

SI personnel are recruited from civilians or from the armed forces of the United States and its Allies. Military personnel of the United States armed forces may be as-

TOP SECRET

signed to OSS within authorized allotment and detailed for SI activities. Military personnel of Allied armed forces may be attached to OSS for SI activities by agreement with authorities of the nations concerned.

10. *TYPES OF PERSONNEL REQUIRED*

  a. The qualifications for SI personnel vary according to the requirements of the assignment. Members of OSS, Washington, or Field Base staffs are selected for integrity, intelligence, initiative, and for special qualifications fitting them for a particular job.

  b. Undercover field operatives are, if possible, selected with specific cover jobs in view. Unless they can conform thoroughly to a reasonable cover, they can not be used. Natural resourcefulness, energy, a broad general background, familiarity with the area, language fluency, patience, discretion, and judgment are important attributes. The operative should be able to handle men, mix easily, judge character so as to be able to deal with agents. Unqualified loyalty to the United States and unqualified sympathy with the vigorous prosecution of the war by all methods are essential.

SECTION IV — TRAINING

11. *ORGANIZATION FOR TRAINING*

  It is the responsibility of the Secret Intelligence Branch to see that proper training is given to its personnel. The Branch's training program is developed by a Training Advisory Committee, a Training Coordinating Committee, and the Branch Training Coordinator. The latter represents SI on the OSS Training Board and handles all relations between SI and the Schools and Training Branch, OSS. Facilities for training at special schools are provided by the Schools and Training Branch, OSS, in the Washington area. The training of agents recruited in the field is the responsibility of OSS field bases and operatives.

12. *SCOPE OF TRAINING*

  a. Before starting formal training, prospective operatives are generally put through a series of tests and observations to determine their aptitudes for SI work.

TOP SECRET

b. Training for operatives begins with a basic course in secret intelligence. This course embraces, both in theory and in practice, such matters as security, cover, communications, recruiting and handling agents, police methods, battle order, effects of propaganda, public opinion testing, cipher, radio code, elementary map reading and sketching, use of the compass, demolitions, and weapons. Advanced training for the secret intelligence operatives is conducted at an SI finishing school. Here specialized instruction is given in secret intelligence techniques, and the operative, with assistance from the instructors and his desk head, develops his cover and otherwise prepares for his particular mission. The desk head can be of great help to the operative during this stage of highly individualized instruction and to that end should remain in as close touch with him as is consistent with security.

c. Until he leaves for the field, the operative continues to receive further special instruction, including special briefing and types of intelligence desired from the area where he will operate. Every assistance possible will be given by his desk head to that end. This training should be given to SI operatives recruited and trained in the U.S. and will be supplemented in the theater. SI operatives and agents selected in the theater will receive similar training and instructions under direction of the SI section chief.

d. New personnel selected, section chiefs, desk heads, and other staff personnel for the SI Branch should take the basic course in order to become familiar with the instruction given at the schools.

13. *TRAINING OBJECTIVES*

a. The SI operative must be able to live according to his cover, often in constant contact with experts in his cover activity, without arousing suspicion. He must know how to employ uncensored or underground methods of communication without undue risk to himself or others. He must be able to report accurately and explicitly and to evaluate information he receives. The ability to evaluate requires not only an under-

TOP SECRET

standing of persons, but familiarity with the military, social, political, economic, and religious history of the area concerned. It requires as well a working knowledge of world-wide current events. The operative going on a mission in an enemy or enemy-occupied country also requires an intimate knowledge of the territory and the ability to pass for a native. Briefly stated, the training of operatives is conducted with these ends in view: to get the operative to his post; to enable him to remain there; to get his reports out; and to get him out when and if necessary.

b. Section chiefs, desk heads, and other staff personnel dealing with operatives require a knowledge of field operating methods and conditions so as better to direct and judge the work of men in the field.

SECTION V — METHODS OF OPERATIONS

14. *GENERAL*

OSS is authorized to conduct secret intelligence activities in all areas, exclusive of the Western Hemisphere. In neutral areas, however, SI activities may be limited by understandings with the chiefs of diplomatic missions.

15. *SECURITY*

a. Security is the *sine qua non* of secret intelligence activities. If security is lacking anywhere in the process of collecting and disseminating information, the continued functioning of an individual or of an entire network is endangered.

b. The factor of security is present to a greater or lesser degree in every phase of secret intelligence activities; however, several general principles governing security may be stated:

(1) No one in a secret intelligence organization should be told more than he has to know to do his own job. The less any one man knows, the less he can let slip — or be forced to tell — if taken by the enemy. As far as possible, the different activities carried on by an intelligence organization should be boxed in water-tight compartments.

TOP SECRET

(2) Secret intelligence personnel should be suspicious of every individual until his loyalty has been proven beyond a doubt.

(3) Secret intelligence personnel should proceed on the assumption that all telephones are tapped, all mail censored, all rooms wired, all radio messages read by the enemy.

c. Following are some of the specific security measures that may be taken in the field:

(1) Cutouts should be used by the operative whenever he considers it unsafe to come into direct contact with another individual.

(2) Meeting places should be selected for the opportunities they afford the participants for an inconspicuous encounter.

(3) Danger signals should be arranged in advance of meetings. In order to avoid detection at the time of signaling, a system should be used whereby a pre-determined signal is given only when it is desired to indicate the absence of danger. If danger is present, no signal will be given.

(4) Recognition signals to be used between persons meeting for the first time should also be arranged in advance.

16. *COVER*

a. Every SI operative and agent working in enemy, enemy-occupied or neutral territory must have a suitable cover — that is, an ostensibly legitimate reason for being where he is.

b. Obviously, cover must be safe. That is, it must successfully shield the operative's secret activities. In the second place, it must allow the operative sufficient freedom of action to perform his mission. For the activities of the operative must be consistent with his cover. The following broad principles govern the selection of cover:

(1) *Social freedom*

A good cover will permit the operative to mingle inconspicuously with the kind of people he will have to see to do his job. His particular mission might require the operative to meet people of all levels of society, in which case his cover should justify such varied association. It is generally easier for a man to associate with those beneath him in the social scale than with those above him. Thus, a doctor or a lawyer can legitimately meet all kinds of people, while it would be suspicious for a stevedore to associate with people in high places. However, some jobs, such as those of waiter or cab driver, allow considerable social freedom and provide effective cover for agents.

(2) *Financial freedom*

A good cover will permit the operative to handle the sums of money his SI activities will require, for he must live within the limits of the income received from his cover occupation. If he is to handle substantial amounts of money and to entertain a good deal, the operative should adopt a cover occupation that pays well. On the other hand, if circumstances require him to adopt a poorly-paid occupation, he must be careful not to spend more money than the income from such an occupation would normally allow. Many covers are wrecked on the rock of finances. Unusual bank deposits or irregular financial transactions are prime causes of counter-espionage investigations.

(3) *Freedom of movement*

A good cover will permit the operative to travel to the extent necessitated by his mission. If his particular mission requires extensive traveling, he should choose a cover that would make frequent journeys perfectly natural. It must be remembered, however, that every trip made must have its particular cover story — a story consistent either with the operative's assumed occupation or with his assumed personal life. This story should be prepared in advance and be as true as circumstances permit.

(4) *Freedom of leisure*

A good cover will allow the operative sufficient leisure time for the conduct of his SI activities.

TOP SECRET

Therefore his cover occupation must not demand too much of his time. If possible, the cover chosen should permit short or irregular hours of work.

$\underline{c}$. In the selection of cover, an occupation should be chosen with which the operative is familiar and which is consistent with his own experience. He should draw as much from his own life as is safe to do. Thus his story will be better able to stand investigation. The most effective cover is that which is as near truth as possible. In any case, the cover selected will be limited by the operative's personal characteristics and abilities, as well as by his mission.

$\underline{d}$. Where the cover is almost, or wholly, artificial, the operative must take every precaution to live the part. His dress, appearance, personal effects, speech, mannerisms, and every action must conform. He must be sure that nothing he wears, possesses, says, or does will make him conspicuous or reveal that he is not what he pretends to be.

$\underline{e}$. Cover is so important, and good covers so rare, that in many cases the finding of a good cover will determine the selection of the operative and the definition of his mission.

$\underline{f}$. The selection of a suitable cover is the responsibility of the section chief or desk head. Arrangements with organizations outside of OSS, either private or governmental, which cooperate in providing cover for an operative, are made through the intermediary of a representative of the Director, OSS, appointed for the purpose.

g. In working out the details of an operative's cover, the desk head will have the assistance of the Document Intelligence Division of the Censorship and Documents (CD) Branch. From this Division, the desk head will be able to obtain for his operative the necessary samples of foreign papers, stamps, labels, letterheads, and documents; required items of foreign clothing, accessories, suitcases, dispatch cases, and similar equipment; and information on conditions and regulations in foreign countries with which the operative must be familiar.

TOP SECRET

h. In the event of capture by the enemy, a secret intelligence operative or agent should stick by his cover story and deny all charges. Despite the seriousness of his own position, he should not fail to protect to the end the security of the organization of which he is a member.

17. *COMMUNICATIONS*

a. Good communications are essential to the efficient functioning of an intelligence network. An operative may be able to obtain vital information, but unless he can get that information to the right people in sufficient time, his work will have been wasted. Much thought and effort, therefore, must go into the establishment of a safe, rapid communications system.

b. Communications can be divided into three categories: within a network; between operatives or agents and the field base; and from a field base to other field bases and Washington.

c. Within a network a number of varying methods may be used to maintain communications. These include personal meetings, cutouts, secret inks, improvised codes, and letter drops, and at times telephone, telegraph, ordinary mail, or general delivery. Each of these measures has particular advantages and disadvantages, and each requires special precautions. The method or combination of methods used will be governed by local conditions. If possible, an alternate communications system should be set up and held in readiness to be used if the first system should break down.

d. For communications between a network and a field base, radio is one of the best means in view of its rapidity. When used, adequate security must be taken to avoid enemy detection. Security methods include: keeping the transmission short; changing the transmission time constantly; moving the location of the set frequently; employing cipher. In addition to radio, couriers are a primary means of communication. Sometimes, however, communication can be effected through

TOP SECRET

transport workers, public conveyances, or even more ordinary methods of telephone, telegraph, or mail.

e. In communicating between a field base and other field bases or Washington, existing Army and Navy, State Department, and commercial facilities will be used.

18. *ARRIVAL AND DEPARTURE*

a. It is essential that careful planning precede an operative's penetration of a new territory and that he be furnished with detailed instructions as to the means of entry and of contacting individuals who will be of assistance to him. This is particularly true of an operative inaugurating SI activities in an enemy or enemy-controlled country.

b. An operative can enter and leave his assigned area of operations either secretly or by the normal means of access and egress under the protection of his cover. An operative may gain secret entry to a territory by airplane, submarine or other vessel, or by making his way across a land border. Particular care must be taken to hide or destroy the paraphernalia an operative may have used to enter a country surreptitiously, such as a parachute or a rubber boat.

c. On arrival in a new area, the operative should learn all he can as quickly but discreetly as possible about local conditions and regulations and local personalities, and should at once plan and make arrangements for his escape in case of emergency.

d. Before he enters a new area, every effort is made to furnish the operative with authentic and current documents, such as identity and ration cards. However, since the enemy authorities may from time to time make changes in the cards currently in effect as a control measure, the operative working in hostile territory should as soon as possible make sure that his documents conform to existing regulations.

e. His first pre-occupation should be to establish himself in his cover and become an accepted member of the community. He should not attempt any undercover work until this preliminary adjustment has been accomplished. The time required to establish himself will depend on where the operative is located, the nature of his cover, his own resourcefulness and the amount of assistance he will receive from friends. Generally speaking, the operative will be able to begin functioning a good deal sooner in a neutral country than in enemy or enemy-occupied territory, where greater precautions must be taken. The operative or agent who is a citizen or resident of the area in which he is to operate has a distinct advantage and will be able to begin his undercover work much sooner.

19. *ESTABLISHING SOURCES OF INFORMATION*

a. In neutral countries, local American business men or those of a friendly nationality can be useful to the operative in making contacts and securing sources of information. Members of the neutral country's secret police and minor government officials, if favorably disposed or sufficiently rewarded, can also be of great assistance. Undercover activities in a neutral country are usually in violation of the laws of the country. Hence, in every case proper security measures must be taken, as well as every precaution against enemy agents in the same area.

b. In enemy and enemy-occupied countries, the operative may receive support and assistance from members of underground organizations and opposition political parties with whom he has established contacts.

c. In the selection of agents, those shall be sought who have direct access to the information desired; first hand information will be more accurate and helpful than hearsay.

d. The number and type of agents an SI operative should recruit will vary with existing local conditions. In general, a secret intelligence network should be kept as small and compact as the mission to be accomplished will allow.

TOP SECRET

e. In conformity with the basic rule of security that no one in the organization be told more than he needs to know to do his own job, a secret intelligence network may be set up along the lines of the cell system, modified to fit prevailing circumstances. The following diagram represents a type of the cell principle:

HEADQUARTERS

```
                    Operative
           /           |           \
       Agent           |           Agent
                       |
                     Agent
                                              Cell No. 1
        /           |           |           \
  Sub-agent    Sub-agent    Sub-agent    Sub-agent
                       |
                   Sub-agent
                   (chief)
                                              Cell No. 2
        /           |           |           \
  Sub-agent    Sub-agent    Sub-agent    Sub-agent
                       |
                   Sub-agent
                   (chief)
                                              Cell No. 3
        /           |           |           \
  Sub-agent    Sub-agent    Sub-agent    Sub-agent
```

TOP SECRET

A key man, the operative, would be sent by headquarters to organize a network in a given area. This man would recruit locally one or more agents, none of whom — in case there were more than one — would know the others. The operative alone would communicate directly with headquarters, each of his agents reporting to him. Each agent would then organize a cell, or group, of perhaps four to six sub-agents. These sub-agents would not know the operative, but each would report to the agent in charge of his particular cell. One of the sub-agents from cell number one, selected for his leadership and ability, would then be designated to form and become chief of a second cell. None of the men in cell number two would know the men in cell number one except the chief, who would report to the head of the first cell. One man from cell number two would then be designated to form a third cell, and so on until the desired number of cells was organized. That number would vary with the job to be done and the local situation. As used by SI, this form of cell organization is not rigid, but may be altered to meet special conditions.

f. Before employing a new agent, the operative should conduct a thorough yet unobtrusive security check to make sure of his reliability. The importance of knowing one's man is obvious. Operatives should be particularly wary of individuals who offer their services unsolicited; these may be agents provocateur and operating for the enemy police. After checking a prospective agent for security, the operative should assign him relatively simple tasks at first, gradually building up to more difficult tasks of greater trust.

g. In addition to collecting information through regularly employed agents, the operative even in enemy or enemy-occupied countries will be able to gather a good deal of general information from the press and radio and through his normal social and business contacts. The individuals furnishing this information, of course, will not be aware of the operative's secret activities. When pieced together in the light of reports received from agents, information gleaned in this manner can prove of value.

20. *REMUNERATION OF AGENTS*

a. Whenever possible, agents should be recruited whose motives for working against the enemy are patriotic rather than financial. Remuneration should be regarded by the agent as a reward rather than as an inducement to render services. Many individuals, particularly in enemy-occupied territory, will serve as agents out of reasons of patriotism. However, agents who volunteer their services should, if willing, be reimbursed for expenses incurred in the performance of SI duties.

b. Relations with agents working for monetary gain should be placed on a business-like basis from the outset. A definite rate of compensation should be agreed upon and adhered to. Some agents work better when they receive a small retainer and are paid over and above this fee according to results. In such instances, a close watch must be kept to check faked reports submitted in order to get easy money. A good man should be treated fairly and generously and appreciation shown for good work.

c. The terms of employment of agents should be reviewed from time to time in the light of results achieved. However, a proven man ought not to be unnecessarily harried the moment he ceases producing results. Every agent experiences unproductive lulls when information is not easily available or is simply non-existent. He should be allowed to feel that he is trusted and that he can expect reasonable certainty of employment in return for his loyal services.

d. In addition to money, remuneration may be in kind — such as food, medicines, clothing — or in services and favors of one sort or another.

e. If agents get into trouble as a result of their SI activities they and their families should be given every aid possible within the bounds of security. In the event

TOP SECRET

an agent is in danger of being uncovered, he should be gotten to safety and provision made for the family he may leave behind.

21. *SPECIALIZATION OF AGENTS*

In ability, training, and cover an agent should be a specialist for the type of intelligence work he is to do. Assigning an agent to a number of different types of intelligence activities reduces his effectiveness and increases the risk of his being uncovered. However, specialization should be kept within limits and not carried to a point where it requires the employment of too many agents.

22. *DOUBLE AGENTS*

a. Double agents — agents working simultaneously for both sides — can prove useful to the SI operative, but the use of double agents is extremely dangerous.

b. There are two general categories into which double agents fall. First is the agent who works one side against the other entirely for personal gain and who does not necessarily know that the operative, his superior, is aware of his double connection. Such an agent can be of limited value and is dangerous to handle. But if employed, particular care must be taken not to divulge, or make it possible for the agent to obtain, any information which would be of use to the enemy.

c. On the other hand, there is the agent whose loyalty to the allied cause is unquestioned and who at the same time has been able to work his way into an enemy intelligence organization. Such an agent is of greater value than the first one; for not only can he supply the enemy with false information, but he may also be able to secure information about the enemy from the inside with some degree of reliability. However, all information obtained from double agents must be checked with extra precaution to guard against deception.

TOP SECRET

23. SEVERING CONNECTIONS WITH AGENTS

If it ever becomes necessary to discontinue the services of an agent, the particular circumstances surrounding the case will govern the procedure to be followed. As a guiding principle, however, it must be remembered that the welfare of the organization comes before that of any individual in it. The operative must be objective, and, if necessary, ruthless in deciding how to handle an agent with whom connections must be severed. However, if secret intelligence activities have been established and conducted with proper regard to security, the services of an agent may be dispensed with without jeopardizing the entire organization.

24. RELATIONS WITH UNDERGROUND GROUPS

a. Underground organizations in enemy or enemy-occupied territory can be of invaluable assistance to the SI operative. They may be able to advise him of the dependability of certain persons, to inform him of the counter-espionage methods of the local authorities, to furnish him with communications, or to help him collect the information he has been sent to obtain. On the other hand, the SI operative, in cooperation with other OSS branches, may be in a position to help the underground. He may be able to provide money, supplies, medicines, communications, and other services which the organization may require to carry on its work.

b. In addition to maintaining relations with underground organizations in the actual area of operations, he should assist in establishing liaison between that organization and the nearest SS field base. In the case of regions not yet penetrated by SI operatives, this liaison may constitute one of the most important sources of information about the region in question.

25. ASCERTAINING PUBLIC OPINION

In determining popular attitudes regarding current issues, every effort should be made to get the opinions of a representative cross-section of the population. Ideally, the various groupings into which the population can be broken down as to age, sex, political party, religion, occupation, income, etc., should be represented proportionately in the group upon whose opinions the operative bases his reports. Such ideal conditions are, of course, difficult for an undercover operative to achieve. His particular cover may enable him safely to associate only with people of a certain level of society. To get the broader picture, however, he should make use of his agents and informants who may be able to move in circles closed to him. In any case, the operative should indicate in his reports on public opinion the extent to which the popular attitudes reported are representative of the people as a whole or of one or more particular groups.

26. *PENETRATING AN ENEMY OR ENEMY-OCCUPIED COUNTRY FROM NEUTRAL TERRITORY*

a. In many cases, the best means of establishing an intelligence network in enemy or enemy-occupied territory is by working through a nearby neutral country.

b. Having established himself under cover in the selected neutral country, the SI operative should make contact with persons who are willing or who can be induced to work for him. The operative will choose individuals — after, of course, taking all necessary security precautions — whose qualifications include the privilege of residing in the enemy or enemy-occupied country and of traveling between it and the neutral country from time to time. Such individuals may be found among minor government officials, political leaders, business men, industrialists, educators, scientists, commercial travelers, seamen, railroad employees and other transport workers. With the aid of their business, social or governmental connections, and under cover of their normal occupations — often as a side-line to those occupations — the individuals thus selected may set up small intelligence networks in the enemy or enemy-occupied country or operate entirely on their own to secure the desired information. They will report to the secret intelligence operative who remains on neutral soil. Professional smugglers may also be employed to get information and materiel into or out of enemy or enemy-occupied countries.

c. If an agent network has been established, it will be easier for the SI operative to enter enemy or enemy-occupied territory to conduct his work personally.

27. *REPORTS*

a. Reports on information collected should be accurate, specific, and timely and should include all pertinent information. Exact dimensions, statistics, and dates should be given whenever possible. Any estimates should be labeled as such. It is essential to indicate the date on which the information contained in the report was observed. Photographs, plans, blueprints, and sketches should be included when they contribute to the clarity of the information.

b. Speed in reporting information is always an important factor. Sometimes it may be the crucial factor which will determine the ultimate success or failure of an operation. Hence every effort must be made to transmit information as rapidly as possible. This calls attention to the need for an adequate communications system (See paragraph 17). Upon the urgency of a particular item of information will depend in part the method of communication used in transmitting it. In an emergency, rapid reporting of the available facts is preferable to delay in the hope of ascertaining the whole story.

However, it must be borne in mind that grave danger to an undercover operative lies in prolonged and regular transmission enabling the enemy to locate the station. Reports, therefore, must be of minimum length, sent from different localities at irregular times.

c. With due regard for its security, the original source of any information reported from the field, should always be indicated for the information of the desk head and the reporting board. Such indication is important because the field base or OSS, Washington, may receive the same information from apparently different sources, whereas in reality various agencies are quoting the same source. If the source is not indicated, this may give rise to the erroneous and possibly dangerous belief that the report has been independently confirmed. Information should be confirmed from other sources wherever possible and such confirmation noted in the report. All information concerning the source such as type of individual, occupation, political prejudices, should be furnished wherever possible but should remain consistent with security considerations.

d. In the dissemination of all SI reports, whether from the field or from Washington, the distribution which has been made should be clearly and completely indicated and reported.

e. A clear distinction should be made in the report between fact, rumor, and opinion.

f. Operational data, as distinguished from intelligence, should be segregated and reported separately.

g. As a guide in reporting military and naval information, the operative will find the Basic Notes of the SI Branch on such subjects as Airfield Reporting, Road Reporting, and Beach Reporting to be useful. These notes are brought up to date from time to time and have been translated into foreign languages according to regional needs. A memorandum prepared by the liaison officer MID-SI/OSS entitled "Requirements of the Military Intelligence Service" should also prove helpful in this respect.

28. *RECORDS AND DOCUMENTS*

Generally speaking, no records should be kept. Extreme caution must be exercised by operatives and agents in the field when it is necessary to be in possession of papers relating to their SI activities. Such papers not only place in jeopardy the safety of the individual in whose possession they may be found, but also may furnish the enemy with sufficient information to cause the liquidation of an entire organization. Thus only papers which are

TOP SECRET

absolutely essential to the functioning of an organization or which are to be sent to the base should be kept; and every possible security measure must be taken to prevent that minimum from falling into enemy hands. The keeping of address books and personnel rosters is particularly dangerous and should not be recorded. In those cases where it is essential to hold documents, the danger factor may be reduced by the use of cipher or of a suitable code disguising the true nature of the information. Incriminating documents of any kind should not be carried on one's person. In that connection, messages between members of an SI organization in areas of operation should be oral whenever possible. If it is necessary to transmit a message or report in writing, the data to be transmitted should be written down only at the last possible moment before its delivery. Documents for transmission to the base should be carefully concealed until they can be safely transmitted.

29. *RELATION BETWEEN SI DESKS, WASHINGTON, AND SI STAFF IN THE FIELD*

a. The Washington desk heads should keep in close touch, through appropriate channels, with the staff at field bases and in neutral areas. They should keep the field heads informed of the Reporting Board's evaluation of field reports, and should transmit to the field information obtained from other sources that will enable the field staff personnel more intelligently to direct their future efforts.

b. At intervals during service in the field, it is desirable for a field desk head and key operatives, when consistent with security and cover, to be brought back to Washington for an exchange of views. In this respect, the governing consideration must be the maintenance of a continuous organization. This will enable them to renew personal contacts with Washington staff members, to be brought up to date on any changes in policies or personnel, and to give the organization the benefit of the intangible aspects of their field experience which are difficult to express in written reports. Conversely,

TOP SECRET

Washington desk heads should, whenever possible, be given the opportunity of visiting the field in order to see conditions at first hand and thus gain experience that will assist them in directing the field work.

c. When a field desk head or key operative returns from the field, the Washington staff should set aside sufficient time for conferences with him. It is advisable for the Washington desk head to make a preparatory review of the field member's problems and reports in advance so as to be thoroughly familiar with them at the time of conferring. Nothing is more damaging to the morale of field personnel than to be given the impression that the Washington staff is too busy to see him and cannot waste much time on him.

30. *RELATIONS BETWEEN FIELD DESK HEADS AND OPERATIVES*

a. The major concern of the field desk head, whether at a field base or in a neutral area, is to keep his network of operatives and agents intact and secure. A desk head should not withdraw an operative from his place of work except under unusual circumstances or except insofar as may be consistent with his cover. His absence and return is likely to be noticed, and endanger his network of agents. The usual conditions which will lead a desk head to make an emergency withdrawal of an operative are signs in the operative of strain which may lead to a breaking point, or evidence that he is in danger of being discovered. Either contingency should be foreseen early enough to permit, whenever possible, his replacement by another operative who can more safely carry on his work.

b. The desk head should maintain close contact with each of his operatives. The morale of an operative will depend upon the degree to which he feels that his desk head is personally and constantly vigilant in directing and caring for him. Before going into the field, the operative should be given a directive that is as specific as possible regarding the types of information required from his area. The success or failure of an operative's

TOP SECRET

mission depends, to a great extent, on the desk head's constant attention to the details of the operative's cover, communications, and movement. Any laxity in these particulars will impair the operative's effectiveness and personal safety.

c. Communication from the desk head to the operative should receive special attention. The operative who hears nothing from his desk head for a considerable period of time is likely to feel that his work is unimportant and unappreciated. The desk head must keep closely in touch with the latest intelligence requirements from the operative's area, and, within limits of security, keep the operative informed of important changes in requirements, and of the value of his work.

SECTION VI — TYPES OF INFORMATION REQUIRED

31. *GENERAL*

Since the SI Branch has the dual function of servicing OSS and disseminating intelligence to a number of widely different organizations, the type of information it is required to collect is extremely varied. The order of priority for the different types of information will depend on the area in question and the prevailing situation, as well as on the needs of the different organizations for the particular area. The following lists, paragraphs 32-36 inclusive, typical secret intelligence objectives.

32. *MILITARY INFORMATION*

a. ORDER OF BATTLE

Identification, location, strength and movement of enemy troop units, aircraft, materiel, and base supply depots.

b. DEFENSIVE POSITIONS

Gun emplacements (number, type, size, range, and condition of guns). AA defenses; fortifications, block-houses, pill boxes, trenches, and

TOP SECRET

barbed wire entanglements; AT defenses; tank traps, road blocks, land mines; communication and transportation facilities.

c. LANDING BEACHES AND BEACH DEFENSES

Location; length, width, slope and composition; depth of water off shore, shoals, reefs; currents, tide and surf conditions; terrains behind beaches; roads leading from beaches to interior.

d. AIRFIELDS

Location; adjacent topography including landmarks; dimensions; length and position of runways; surface; size of plane capable of using; obstructions near field; hangars; dispersal areas; repair facilities; fuel and oil supplies; communication and transportation facilities; connecting roads and railroads; defenses; camouflage; weather conditions.

e. COMMUNICATIONS

Railroads, roads, waterways, harbors, radio nets, telephone and telegraph systems used by the military. (See paragraphs 34 d, e, f, g, h, i, below for specific items of information required.)

f. RADAR EQUIPMENT

Type; location; how defended; how camouflaged.

g. SECRET WEAPONS AND LATEST TECHNICAL DEVELOPMENTS.

33. *NAVAL INFORMATION*

a. NAVAL SEA AND AIR FORCES

Location; numbers and identification of vessels by types and names; new or unusual types of vessels; secret weapons and devices; movements of vessels; photographs and silhouettes of vessel and aircraft.

TOP SECRET

b. NAVAL BASES

Location; number and types of vessels present; facilities for construction and repair of vessels, including dry docks; fuel storage and facilities; supply depots; ammunition depots; personnel depots; berthing piers; anchorage ground; air bases; radio stations; radar installations; defenses, land and air, submarine nets and mines.

34. *ECONOMIC INFORMATION*

a. ESSENTIAL WAR INDUSTRIES

Location of plants; type, quantity and quality of production; defenses; camouflage; movement of plants to new locations; effects of bombing; blueprints and plans; sources of supply; labor relations.

b. ELECTRICAL POWER INSTALLATIONS

Location, capacity and defenses of power stations, power dams and high tension lines.

c. TELEPHONE AND TELEGRAPH SYSTEMS

Communication net; location of exchanges; number of wires or cables above or below ground; type and condition of equipment; telephone directories; measures of defense.

d. RADIO COMMUNICATIONS

Location, power, range, wave length and call letters of stations; number, type and condition of receiving sets; best times for reception; measures of defense.

e. RAILROADS

Location; motive power, steam or electric; if the latter, location of controls, transformers, and substations; signal system; number of tracks; gauge; condition of road-bed, rolling stock and equipment; location of tunnels, bridges, culverts, whether prepared for destruction; repair shops, stations, marshalling yards, sidings, switches and turntables; goods and personnel transported; time tables; measures of defense.

f. ROADS

Location; width and surface; viaducts, bridges and culverts, with load capacity; road-building equipment; defenses; trucks; buses and cars available; filling stations, gasoline and oil stocks; amount and kind of traffic.

g. WATERWAYS

Location; width, depth; locks; bridges; barges; defenses; amount and kind of traffic.

h. HARBORS

Number and size of port facilities; transportation and communication facilities; cranes; storage, refrigeration facilities; fuel facilities; fresh water supplies; labor conditions; measures for defense; number and types of vessels, with destination where possible.

i. In general, all important economic changes, such as: marked shortages; greatly increased production; new factories, transportation and communication facilities; destruction by bombing; repairs to installations damaged by bombing; new defense measures. In addition, SI agents may be called upon for specific information on designated areas, such as water supply and health conditions.

35. *POLITICAL INFORMATION*

a. PROPOSED CHANGES IN GOVERNMENT POLICY

b. POLITICAL PARTIES

Aims, strength, importance.

c. POLITICAL PERSONALITIES

Venalities, weaknesses, comprising activities; strong and weak points in ability and character.

TOP SECRET

<u>d</u>. UNDERGROUND MOVEMENTS

Organization; strength; aims; operations; finances; leaders.

<u>e</u>. LABOR ORGANIZATION

Strength; leaders; policies.

<u>f</u>. POLICE SYSTEM

Organization; methods; important officials.

<u>g</u>. HOSTILE INTELLIGENCE AND COUNTER-INTELLIGENCE SERVICES

Organization; aims; methods; relation to similar organizations in other countries; effectiveness; descriptions and personal histories of officials and agents.
(The collection of such information is primarily the responsibility of X-2. However, any information of this nature collected by SI will be passed on to X-2.)

<u>h</u>. METHODS OF CONTROLLING CIVILIAN POPULATION

Identity cards; curfews; travel permits; rationing and other regulations; plans for civilian control on D-day. Copies of identity cards and similar documents should be procured, together with intelligence for their proper use, to be utilized by future agents.

<u>i</u>. Where applicable, relations between enemy occupation authorities and civil population; between enemy occupation authorities and local government; between enemy occupation authorities and local police.

36. *PSYCHOLOGICAL INFORMATION*

<u>a</u>. MORALE OF CIVILIAN POPULATION

War workers, foreign labor, miners, farmers, civil servants, etc.

<u>b</u>. MORALE OF THE ARMED FORCES

Relations between officers and men, between

TOP SECRET

various services, between allied enemy troops, between troops and conquered peoples, between military and civilians; effect on morale of pay, food, housing, medical care, equipment, leaves, etc.; discipline; military smartness; sale of equipment.

c. MORALE EFFECTS OF BOMBING

d. POPULAR ATTITUDES TOWARD THE GOVERNMENT, THE ARMED FORCES, THE CHURCH, UNITED NATIONS, AXIS COUNTRIES

e. EFFECTS OF UNITED NATIONS' PW ON MORALE

f. ENEMY PW METHODS AND RESULTS

g. CLEAVAGES BETWEEN GROUPS OF THE CIVIL POPULATION AND BETWEEN IMPORTANT ELEMENTS OF MILITARY AND GOVERNMENTAL PERSONNEL

SECTION VII — COOPERATION OF SI WITHIN OSS AND WITH OTHER ORGANIZATIONS

37. *GENERAL*

The very nature of its activities requires the SI Branch to coordinate such activities with the other branches of OSS, with the armed services, with the State Department and with the secret intelligence services of the allied nations. Not only must SI be kept informed of the specific intelligence requirements of each agency it serves; but it must see to it that the desired intelligence, once collected and evaluated, is disseminated to the agencies which can act on the basis of that intelligence. Furthermore, duplication of effort will be avoided and the collection and evaluation of information facilitated by a complete, rapid interchange of information with other intelligence organizations.

38. *COOPERATION WITH OTHER BRANCHES OF OSS*

Cooperation of SI with other branches of OSS falls into two general categories:

a. COOPERATION WITHIN THE INTELLIGENCE SERVICES

The activities of the Secret Intelligence, Counterespionage (X-2), Foreign Nationalities (FN), Research and Analysis (R&A) and CD Branches of OSS are coordinated by the Deputy Director, Intelligence Services. Regular meetings of the chiefs of these branches are held under his supervision. Close collaboration must exist between SI and X-2. In many cases SI and X-2 have to operate jointly, with the personnel of both branches being used interchangeably to perform SI and X-2 missions. Close and constant collaboration must also exist between SI and R&A, both in Washington and in the field. In order to facilitate the work of both branches, corresponding desks and section chiefs in SI and R&A acquaint themselves generally with the types of activities (other than purely operational data) currently being undertaken by each other. Foreign Nationalities can serve SI with respect to recruitment.

b. COOPERATION WITH THE OPERATIONS BRANCHES

SI, together with the other intelligence branches of OSS, furnishes the Operational Groups, Maritime Unit, Special Operations and Morals Operations Branches with information which these branches require to plan and execute their missions. On the other hand, in the course of operations, OG, MU, SO, and MO may uncover valuable information which they will pass on to SI for evaluation and dissemination to other interested organizations.

39. *COOPERATION WITH THE ARMED FORCES*

a. It is essential that SI cooperate closely with the armed forces both in Washington and in theaters of operations.

b. In Washington, liaison is maintained between the Military Intelligence Division (MID) and SI on a reciprocal basis in order to insure a free and rapid interchange of appropriate intelligence. Similar arrangements exist with the Office of Naval Intelligence (ONI) and Air Intelligence (A-2).

TOP SECRET

<u>c</u>. In theaters of operation, the coordination of SI activities with the armed forces is still more complete, since all OSS operations come under the direct control of the theater commander.

<u>d</u>. Although combat intelligence does not normally come within the province of SI, SI organizations in theaters of operations may at times be called upon to assist in the collection of this type of intelligence. However, SI should not engage in the collection of combat intelligence unless specifically requested to do so by the theater commander.

40. *COOPERATION WITH THE DEPARTMENT OF STATE*

<u>a</u>. The SI Branch furnishes the State Department with political and other pertinent intelligence, supplementing the intelligence which the Department gathers through its own sources.

<u>b</u>. On its part, the State Department often provides valuable assistance and advice for SI operatives and agents. In addition, diplomatic and consular officials, because of their experience in the field and familiarity with the local scene, are often valuable counselors.

41. *COOPERATION WITH SIMILAR AGENCIES OF ALLIED NATIONS*

The SI Branch maintains close relations with the secret intelligence organizations of allied nations, including those governments-in-exile. These organizations often are able to supply SI with information and personnel which it would be otherwise difficult or impossible to obtain. Close liaison with these organizations prevents duplication of effort and helps provide a check on information secured through other channels. This liaison is maintained both in Washington and at field bases, principally the latter.

TOP SECRET

SECTION VIII — PLANNING

42. *GENERAL*

a. Planning for secret intelligence is essential. Detailed planning must be performed before undertaking activities in a given area and before extending activities already under way.

b. After an organization is established in a given area, secret intelligence activities are determined largely in accordance with the information demanded by a changing situation. Thus the planning of those activities must be a continuous process, based upon military and political developments. It must be flexible and should take advantage of opportunities as they present themselves. It must be farsighted. It will be dependent in large measures upon the initiative of the personnel of all echelons.

43. *PROGRAMS*

a. Special Programs covering SI activities in a theater of operations are incorporated into OSS Over-All Programs. In the Over-All Program for a given theater or area, the objectives for all the OSS branches concerned are set forth in order of importance. The Special SI Programs state the missions to be performed by SI to attain the general objectives listed in the Over-All Program, present a brief summary of the situation bearing on the missions in question and prescribe in a general way the plan to be followed. These special SI Programs are drawn up jointly by the Strategic Services Planning Staff and the SI Branch, and are presented to the OSS Planning Group for approval. Upon approval by the Planning Group, the Programs are submitted to the Director, OSS for his consideration and approval before being transmitted to OSS, field bases or to OSS missions in neutral areas.

b. Subject to the approval of the theater commander, SI Programs establish priorities for SI activities in the field. In conformity with these special programs, SI pre-

TOP SECRET

pares detailed operational plans.

<u>c</u>. In Washington detailed operational plans and projects in implementation of special programs are drawn up by the SI desk head in consultation with the section chief. These plans and projects are then routed through the SI Projects Officer, for recording, to the Chief of the SI Branch, for approval. The Planning Officer, SI, coordinates the plans originating in any one section with other plans of the SI Branch. At the same time he is responsible for coordinating plans of the SI Branch with those of other branches of OSS.

<u>d</u>. When SI, Washington, or at the field base draws up plans or projects unrelated to the objectives and missions covered in a special program, such plans and projects are reported to the Director, Washington, attention Planning Group, for approval and incorporation into the appropriate program.

<u>e</u>. The Chief of SI Section, OSS Field Bases, and in neutral countries, will provide the Strategic Services Officer of the Chief of OSS Mission with all necessary information on progress of activities under SI Special Programs for inclusion in his regular report to OSS, Washington.

44. *IMPLEMENTATION STUDIES*

SI, as well as the other OSS branches concerned, participates in the preparation of Implementation Studies. These studies support the various Over-All and Special Programs. They cover certain specific areas and provide supporting data and background information to be used in the preparation of operational plans.

45. *CHECK LIST*

In Appendix "A" there are summarized in the form of a check list a number of the more important points that have been presented in this manual. This check list may serve as a brief list of reminders to SI personnel to assist them in the course of their work.

TOP SECRET

APPENDIX "A"
TO
SECRET INTELLIGENCE FIELD MANUAL —
STRATEGIC SERVICES
CHECK LIST FOR SI ACTIVITIES

This check list is designed to assist desk heads, section chiefs, and other staff personnel, in Washington and in the field, in the planning and execution of SI activities.

FOR SI (WASHINGTON)

1. *AUTHORITY*

a. Does the projected activity conform to approved Strategic Services Over-all and Special SI Programs?—or to contemplated additional programs or revisions of such programs?

2. *PLANNING IN IMPLEMENTATION OF SI PROGRAMS*

a. Is planning complete?

b. Has the plan been approved by proper authority?

c. Have provisions been made for:

(1) Recruitment and training of necessary personnel?

(2) Equipment, supplies, funds, and administrative services?

(3) Adequate and secure communications?

(4) Transportation to the theater?

d. Has the advisability been considered of coordinating SI plans with those of other OSS branches and appropriate Allied agencies?

e. Do plans provide that all pertinent intelligence is available in the field for use in indoctrination of personnel for implementing this plan?

_f_. Have arrangements been made to send to the field:

(1) Comments on data received and further data desired?

(2) Evaluation on the operatives reports?

3. *PERSONNEL*

_a_. Have adequate "cover" arrangements been made?

_b_. Have personnel about to be sent abroad in connection with prospective activities been checked individually for:

(1) Proper training?

(2) Required inoculations and physical examinations?

(3) Knowledge of types of information required?

(4) Regular and special equipment?

(5) Security?

_c_. Has proper security examination been made of all personnel to make certain they understand all security provisions?

4. *TRAINING*

_a_. Has contact been kept by the desk head with men in training?

_b_. Has special training for the specific assignment been completed satisfactorily?

_c_. Has the desk head discussed with each of his men, as completely as is possible, consistent with security, the proposed assignment?

_d_. Has personnel been properly indoctrinated?

_e_. Has sufficient emphasis been placed on security during and on completion of the training course?

_f_. Are you satisfied with the security and discretion of the individual?

TOP SECRET

5. *EQUIPMENT*

<u>a</u>. Has meticulous attention been paid to the equipment of the individual?

(1) Has he been properly equipped as an SI operative prior to departure for overseas?

(2) Has particular attention been given to foreign funds, foreign documents, cover clothing, and communications equipment?

(3) If not available in Washington, are they available in the field?

(4) If not, have you made other arrangements for their procurement?

<u>b</u>. Have arrangements been made with Procurement and Supply to furnish special OSS equipment?

<u>c</u>. Has the base been notified of what part of the supplies not carried with personnel will be sent from Washington?

<u>d</u>. Has branch chief in the field been notified to initiate requests for supplies and equipment as soon as a possible shortage can be foreseen?

6. *SHIPMENT OF SUPPLIES*

<u>a</u>. Has theater commander's approval been received from the field for shipment of supplies and equipment?

<u>b</u>. Has field been informed of:

(1) Schedule of shipment of supplies and equipment?

(2) Shortages in the shipment?

7. *TRANSPORTATION OF PERSONNEL*

<u>a</u>. Has approval of the theater commander been received for transportation of personnel?

<u>b</u>. Have the proper documents been prepared and all authorizations obtained?

TOP SECRET

<u>c</u>. Has "overseas security check" been obtained?

<u>d</u>. Has final inspection been made of physical condition and equipment of personnel?

<u>e</u>. Has the field been notified of the names, grades, and ratings of military personnel being sent, (names only of civilians), as well as of the number that is to follow, if any, to complete the requirements for the projected activity or task?

8. *REPORTS*

<u>a</u>. Are reports on SI activities received from the field?

<u>b</u>. Do reports indicate that the SI activities conform to approved Strategic Services Special SI Programs?

<u>c</u>. Are the reports from the field complete and in the prescribed form?

FOR SI (FIELD)

1. *AUTHORITY*

Does the projected activity conform to approved Strategic Services Special SI Programs or to additional activities approved by competent authority for inclusion in Special Programs?

2. *PLANNING IN IMPLEMENTATION OF SI PROGRAMS*

<u>a</u>. Is planning complete?

<u>b</u>. Has the plan been approved by proper authority?

<u>c</u>. Have provisions been made for?

(1) Recruitment and training of necessary additional personnel in the theater?

(2) Equipment, supplies, funds, and administrative services?

(3) Adequate and secure communications?

TOP SECRET

(4) Transportation to and within the area of operations?

d. Has the advisability been considered of coordinating SI plans with those of other OSS branches, military agencies within the theater, and appropriate Allied agencies?

e. Is plan in accordance with most recent intelligence from OSS and other available sources?

3. *PERSONNEL*

a. Has recruitment in conformity with your plan been initiated in the theater?

b. Has personnel on arrival from overseas been examined individually for:

(1) Morale?

(2) Physical condition?

(3) Equipment?

(4) Training?

(5) Indoctrination?

(6) Security?

c. Have final "cover" details been arranged?

4. *TRAINING*

For personnel trained at the field base, have the following points been checked?

a. Has continuous contact been kept by the desk with men in training?

b. Has special training for the specific assignment been completed satisfactorily?

c. Has the desk head discussed with the men, as completely as is possible consistent with security, the proposed assignment?

d. Is the indoctrination complete?

TOP SECRET

e̲ Has sufficient emphasis been placed with the individual on security during the training course and while waiting departure on his task?

5. *EQUIPMENT*

a̲. For the individual intended to enter enemy or enemy-occupied countries, are you satisfied with his personal equipment, clothing, foreign funds, foreign documents, and communications equipment?

b̲. Have arrangements been made for furnishing future equipment, funds, and documents to the individual?

6. *SHIPMENT OF SUPPLIES*

a̲. Have necessary requisitions been forwarded to OSS, Washington to equip agents recruited and trained in the field?

b̲. Has theater commander approval been forwarded to Washington for shipment of items?

c̲. Has schedule for shipments been worked out with Washington?

7. *INTRODUCTION INTO ENEMY OR ENEMY-OCCUPIED COUNTRIES*

a̲. Have all the proper documents been prepared and all authorizations received from competent authority?

b̲. Has proper liaison been arranged with appropriate Allied agencies or resistance groups?

c̲. Have arrangements been made with X-2 in connection with departure of personnel and its activities in the field?

d̲. Have arrangements been made for withdrawal of the mission when its task is completed?

e̲. Have proper arrangements been made with appropriate Allied agencies, resistance groups, or our own agents assisting in the reception of personnel?

8. *REPORTS*

Have arrangements been made to transmit to Washington reports on:

a. All military, naval, political, economic, and psychological information required?

b. Operational plans made in the implementation of special programs?

c. Successes, failures, and difficulties in accomplishing missions?

d. Effectiveness of OSS equipment and devices?

e. Any new methods developed?

f. Status of personnel — by activities?

g. Cooperation with Allied organizations?

www.ingramcontent.com/pod-product-compliance
Lightning Source LLC
Chambersburg PA
CBHW050245230526
45470CB00005B/2116